EXPLORE!

FORCE OF NATURE

T0414164

TORNADOES

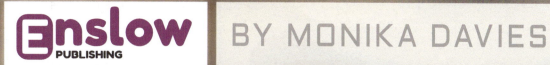

BY MONIKA DAVIES

Please visit our website, www.enslow.com. For a free color catalog of all our high-quality books, call toll free 1-800-398-2504 or fax 1-877-980-4454.

Library of Congress Cataloging-in-Publication Data

Names: Davies, Monika, author.
Title: Tornadoes / Monika Davies.
Description: New York : Enslow Publishing, [2021] | Series: Force of nature
  | Includes bibliographical references and index.
Identifiers: LCCN 2019050669 | ISBN 9781978518490 (library binding) | ISBN
  9781978518483 (paperback) | ISBN 9781978518506 (ebook)
Subjects: LCSH: Tornadoes–Juvenile literature.
Classification: LCC QC955.2 .D383 2021 | DDC 551.55/3–dc23
LC record available at https://lccn.loc.gov/2019050669

Published in 2021 by
Enslow Publishing
101 West 23rd Street, Suite #240
New York, NY 10011

Copyright © 2021 Enslow Publishing

Designer: Katelyn E. Reynolds
Editor: Monika Davies

Photo credits: Cover, p. 1 solarseven/Shutterstock.com; cover, pp. 1–48 (series art) Merfin/Shutterstock.com; p. 5 Willoughby Owen/Moment/Getty Images; p. 6 ohenze/Shutterstock.com; p. 7 bauhaus1000/E+/Getty Images; p. 8 Designua/Shutterstock.com; p. 9 Dan Ross/Shutterstock.com; p. 10 Jason Persoff Stormdoctor/Cultura/Getty Images; p. 11 Rainer Lesniewski/Shutterstock.com; pp. 12, 18 VectorMine/Shutterstock.com; p. 13 Minerva Studio/Shutterstock.com; p. 14 Mike Mareen/Shutterstock.com; p. 15 Arshad876/Shutterstock.com; p. 16 swa182/Shutterstock.com; p. 17 Kent F. Berg/The Miami Herald/Getty Images; p. 19 PeteDraper/E+/Getty Images; pp. 20, 28, 30 Drew Angerer/Getty Images; p. 21 Tasos Katopodis/Getty Images; pp. 22, 26 john finney photography/Moment/Getty Images; pp. 23, 34, 35 Jim Reed/Corbis NX/Getty Images Plus; p. 24 U.S. Navy/Handout/Greg Messier/Getty Images; p. 25 DAVID MCNEW/AFP via Getty Images; p. 27 Bettmann/Getty Images; p. 29 Jason Persoff Stormdoctor/Image Source/Getty Images; p. 31 courtesy of the NOAA; p. 32 Niccolò Ubalducci Photographer–Stormchaser/Moment/Getty Images; p. 37 Topical Press Agency/Hulton Archive/Getty Images; p. 38 SETH HERALD/AFP via Getty Images; p. 39 Benjamin Krain/Getty Images; p. 40 MANDEL NGAN/AFP via Getty Images; p. 41 DigitalGlobe via Getty Images; p. 42 Joe Raedle/Getty Images; p. 43 Photoguru73/Shutterstock.com; p. 44 Mike Hollingshead/Corbis NX/Getty Images Plus; p. 45 Cheryl A. Meyer/Shutterstock.com.

Portions of this work were originally authored by Kristen Rajczak and published as *Terrifying Tornadoes*. All new material in this edition was authored by Monika Davies.

Printed in the United States of America

Some of the images in this book illustrate individuals who are models. The depictions do not imply actual situations or events.

CPSIA compliance information: Batch #BS20ENS: For further information contact Enslow Publishing, New York, New York, at 1-800-542-2595.

Find us on

CONTENTS

 Deadly Twisters 4

 Tornado Fact File 6

 Types of Tornadoes 16

 Tracking Tornadoes 26

 Tornado Training 36

 Glossary 46

 For More Information 47

 Index 48

WORDS IN THE GLOSSARY APPEAR IN **BOLD** TYPE THE FIRST TIME THEY ARE USED IN THE TEXT.

DEADLY TWISTERS

Often seen across a dramatic skyline, a tornado is a whirling and twisting dark column of air that leaves mass **damage** in its wake. Tornadoes are a common sight in many areas of the United States. These forces of nature can pull up large oak trees from their roots, can destroy entire apartment buildings, and are sometimes responsible for a high number of deaths and injuries.

While tornadoes hit many parts of the world, the United States is where the majority of tornadoes strike. In this book, we'll examine how tornadoes develop, the different types of tornadoes, and how we can track and prepare for the tornadoes of the future.

A TORNADO CAN LAST FROM BETWEEN A FEW MINUTES TO A FEW HOURS. HOWEVER, MOST TORNADOES TOUCH THE GROUND FOR ONLY ABOUT FIVE MINUTES.

TORNADO FACT FILE

A tornado is a column of fiercely **rotating** winds that extends from a storm cloud to the ground. In most places in the United States, wind speeds usually don't exceed 15 miles (24 km) per hour. The most powerful tornadoes have winds that reach more than 300 miles (483 km) per hour!

Tornadoes can cause a lot of harm in a community. They rip up trees and knock down buildings. It's important to know how to be safe during a storm like this. First, let's learn how tornadoes form.

A TORNADO WATCH IS ISSUED TO LET CITIZENS KNOW THEIR AREA HAS CONDITIONS THAT COULD CAUSE A TORNADO AND THEY SHOULD BE PREPARED.

EXPLORE
MORE

IN 1950, THE NATIONAL WEATHER SERVICE (NWS) BEGAN STUDYING, FORECASTING, AND WARNING THE PUBLIC WHEN A TORNADO MIGHT HIT. THE NWS ISSUES A TORNADO WARNING WHEN A TORNADO IS EXPECTED AND CITIZENS SHOULD LOOK FOR SAFE SHELTER.

FORECAST:
TO MAKE AN INFORMED GUESS ABOUT FUTURE WEATHER

About 90 percent of tornadoes form from thunderstorms. Thunderstorms commonly occur when a warm **front** meets a cold front or when a wet air mass meets a dry air mass. Both events push warm, wet air high into the atmosphere. There, moisture in the air **condenses** and creates cumulonimbus clouds, which are thick, tall rain clouds. Cooler air moves in below to replace the rising warm air, resulting in wind.

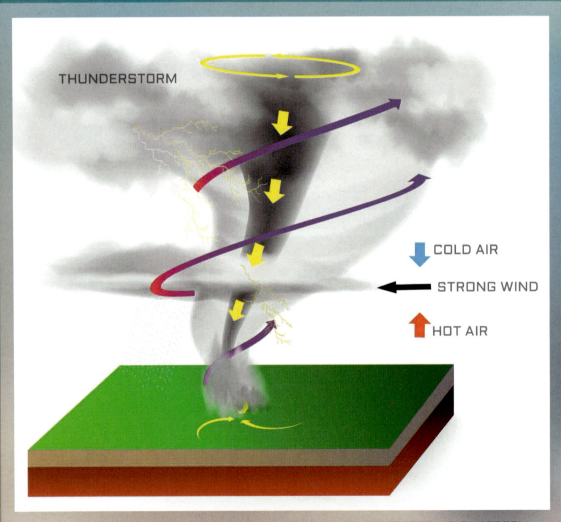

THUNDERSTORM

COLD AIR

STRONG WIND

HOT AIR

MOST TORNADOES FORM FROM SUPERCELL THUNDERSTORMS. SUPERCELL THUNDERSTORMS ARE THOSE THAT LAST LONGER THAN AN HOUR. THEY GROW FROM A LEANING, SPINNING UPDRAFT THAT CAN BE 10 MILES (16 KM) ACROSS AND 50,000 FEET (15,240 M) TALL!

SUPERCELL THUNDERSTORM

EXPLORE MORE

NOT ALL SUPERCELL THUNDERSTORMS LEAD TO A TORNADO. IN FACT, ONLY AROUND 20 PERCENT OF SUPERCELL THUNDERSTORMS ACTUALLY CREATE A TORNADO! SCIENTISTS ARE STILL UNSURE WHY THIS IS THE CASE.

The storm's strength gathers as the cloud mass grows. Water droplets in the clouds have electrical charges that create lightning and the accompanying thunder. When wind speed increases and wind direction changes, a tornado may form.

ALTHOUGH A SUPERCELL THUNDERSTORM IS THE LEAST COMMON TYPE OF THUNDERSTORM, IT IS THE MOST DANGEROUS TYPE.

U.S. TORNADO DATABASE

Tornadoes have happened on every continent except Antarctica. By far, the United States is the country with the most tornadoes every year. Each year, around 1,253 tornadoes occur in the United States. The tornadoes in the United States are also the strongest and most destructive in the world. Tornadoes have occurred in every U.S. state. They're found in the greatest number and strength in the central part of the country, which is nicknamed Tornado Alley for this reason. This area includes parts of Texas, Oklahoma, Kansas, Nebraska, South Dakota, and Colorado.

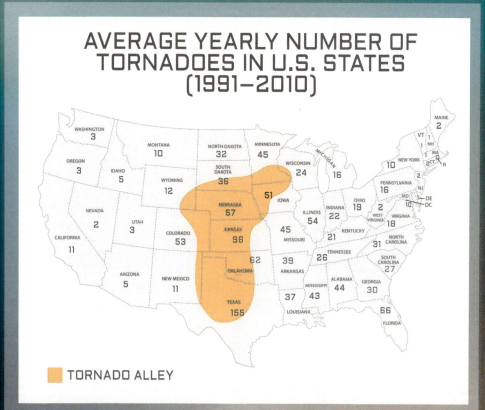

AVERAGE YEARLY NUMBER OF TORNADOES IN U.S. STATES (1991–2010)

- TORNADO ALLEY

THE UNITED STATES HAS THE WORLD'S LARGEST NUMBER OF TORNADOES EVERY YEAR. THE NEXT COUNTRY TO EXPERIENCE A HIGH NUMBER OF TORNADOES IS CANADA, WHICH HAS AROUND 100 TORNADOES EVERY YEAR.

DESTRUCTIVE:
CAUSING SOMETHING TO BE DESTROYED OR RUINED

11

Changes in wind speed and direction create spinning, **horizontal** winds. The column of warm air angles up into the clouds and begins to spin **vertically**. This huge area of rotating air—called a mesocyclone—can be from 2 to 6 miles (3 to 10 km) wide. Within this area, marked by a mass of low-lying clouds called a wall cloud, tornadoes form.

HOW A TORNADO FORMS

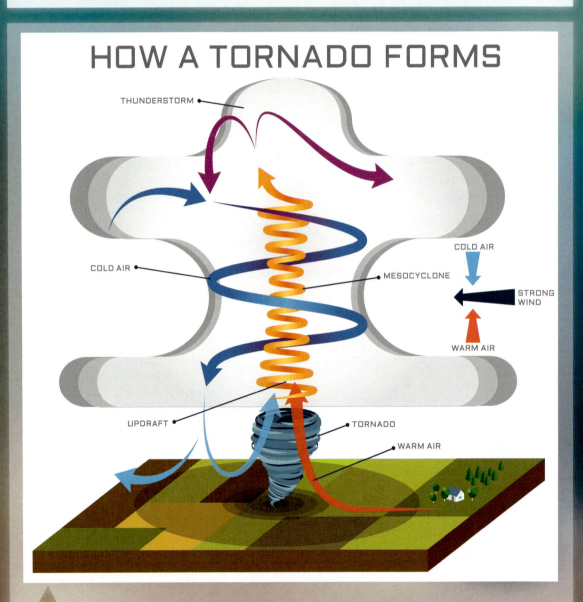

THUNDERSTORM

COLD AIR

COLD AIR

MESOCYCLONE

STRONG WIND

WARM AIR

UPDRAFT

TORNADO

WARM AIR

STORM CHASERS

The 1996 movie *Twister* was about scientists who chase tornadoes to study them. Storm chasers don't just exist in movies. There are real-life storm chasers who put themselves in danger! Some are scientists who want to learn more about how weather happens. Some are photographers who want to take pictures of storm scenes as they happen. However, real-life storm chasers are usually experts who have training. They understand storm conditions, as well as the dangers and risks of chasing a storm. Storm chasers can drive around looking for a tornado for 12 hours or more! Storm chasing often involves a lot of waiting time too.

THE LOOK OF A FUNNEL CLOUD CHANGES AS IT GAINS AND LOSES POWER. SOME TORNADOES CAN'T BE SEEN AT ALL UNTIL THEY PICK UP DUST AND **DEBRIS**!

Tornadoes can also form when a hurricane comes ashore. The powerful winds and great masses of moist air that hurricanes transport can create tornado conditions. Many states along the Gulf of Mexico experience tornadoes just before a hurricane reaches land.

HURRICANE:
A POWERFUL STORM THAT FORMS OVER WATER AND CAUSES HEAVY RAINFALL AND HIGH WINDS

14

TORNADOES THAT FORM FROM HURRICANES ARE COMMONLY LESS POWERFUL THAN THE TORNADOES THAT FORM FROM SUPERCELL THUNDERSTORMS.

WEAK TO VIOLENT

WEAK TORNADOES
- ABOUT 60–70 PERCENT OF ALL TORNADOES
- WINDS LESS THAN 110 MILES (177 KM) PER HOUR

STRONG TORNADOES
- ABOUT 35 PERCENT OF ALL TORNADOES
- WINDS 110 TO 165 MILES (177 TO 266 KM) PER HOUR

VIOLENT TORNADOES
- ABOUT 2 PERCENT OF ALL TORNADOES
- WINDS MORE THAN 166 MILES (267 KM) PER HOUR

TYPES OF TORNADOES

Do you picture a dark, swirling column of wind when you think of a tornado? This column is called a funnel cloud when it isn't in contact with the ground. At this point, it may also be called a condensation funnel. Usually, the more water vapor there is in the air and the stronger the winds are, the larger the column will be. It isn't officially called a tornado until it touches the ground.

16

IN THE EARLY MONTHS OF 2019, MANY U.S. STATES HAD MORE TORNADOES THAN EXPECTED, INCLUDING TEXAS AND MISSISSIPPI.

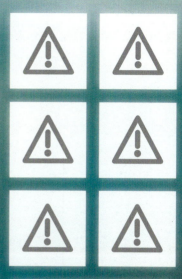

EXPLORE MORE

IN 2010, MINNESOTA HAD MORE TORNADOES THAN ANY OTHER STATE. OF THE MORE THAN 113 TORNADOES THAT HIT MINNESOTA IN 2010, 48 TOUCHED DOWN ON A SINGLE DAY—JUNE 17. BEFORE THIS, MINNESOTA'S RECORD FOR MOST TORNADOES IN ONE DAY HAD BEEN 27, SET IN 1992.

THIS IMAGE IS FROM A 1997 TORNADO THAT TOUCHED DOWN IN MIAMI, FLORIDA. FLORIDA HAS MANY TORNADOES BECAUSE OF THE STATE'S HIGH NUMBER OF THUNDERSTORMS. HOWEVER, TORNADOES IN FLORIDA ARE GENERALLY WEAKER THAN MOST OTHER TORNADOES.

Earth's rotation is partly responsible for the way tornadoes turn. Tornadoes are usually cyclonic. This means they turn counterclockwise in the Northern Hemisphere and clockwise in the Southern Hemisphere. Less than 2 percent of tornadoes are anticyclonic. That means they spin in the opposite direction. Anticyclonic tornadoes form in areas with cool, sinking air.

HEMISPHERE:
ONE-HALF OF EARTH

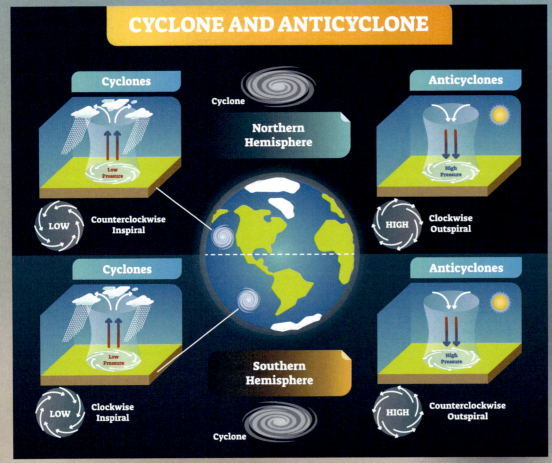

CYCLONE AND ANTICYCLONE

Cyclones

Cyclone

Northern Hemisphere

Low Pressure

LOW Counterclockwise Inspiral

Anticyclones

High Pressure

HIGH Clockwise Outspiral

Cyclones

Low Pressure

LOW Clockwise Inspiral

Cyclone

Southern Hemisphere

Anticyclones

High Pressure

HIGH Counterclockwise Outspiral

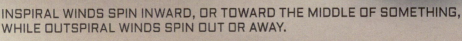

INSPIRAL WINDS SPIN INWARD, OR TOWARD THE MIDDLE OF SOMETHING, WHILE OUTSPIRAL WINDS SPIN OUT OR AWAY.

DAMAGING EFFECTS

Tornadoes can cause mass destruction. Buildings often fall during a tornado. A tornado's winds move at high speeds. These wind speeds can reach as high as 300 miles (483 km) per hour! The tornado's fast winds move over a building and pull up. At the same time, the rapidly moving air rushes around the building's corners and pulls out. When the windows and doors break, air rushes in, pushing the walls and roof up. This can happen so quickly it looks as if the building has exploded! A tornado can also pull cars and trucks off the ground, damage bridges, and flip over trains.

IN THE UNITED STATES, TORNADOES ARE RESPONSIBLE FOR CAUSING AROUND $400 MILLION IN DAMAGE EVERY YEAR.

There are several types of tornadoes. Supercell tornadoes form from supercell thunderstorms. These thunderstorms may produce wedge-shaped tornadoes with violent winds that can be more than 200 miles (322 km) per hour. Supercell tornadoes are the type most likely to remain in contact with the ground for an hour or more.

THIS PHOTOGRAPH FROM MAY 2017 SHOWS A SUPERCELL THUNDERSTORM BUILDING IN QUANAH, TEXAS.

LONG-TRACK TORNADOES

Sometimes, tornadoes last for hours and cause long paths of damage. These are called long-track tornadoes. Some scientists define a long-track tornado as having a path of damage measuring longer than 100 miles (161 km), while a very-long-track tornado has a path of damage longer than 150 miles (241 km). However, it's hard to prove whether a damage path belongs to just one powerful tornado or a tornado family. These groups usually have two or three tornadoes that form from the same central rotating wind.

THIS PATH OF TORNADO DESTRUCTION OCCURRED IN WASHINGTON, ILLINOIS, IN 2013.

VIOLENT:
VERY FORCEFUL

A landspout is a non-supercell tornado. Landspouts aren't as strong as supercell tornadoes. They form from cool, wet air traveling with a thunderstorm. When a landspout's air column touches the ground, it sucks up a layer of dust, giving it another name—dust-tube tornado.

A landspout is similar to a waterspout, or a tornado that happens over water. Waterspouts can form from supercell thunderstorms, but they're small and weaken as soon as they hit land.

A LANDSPOUT USUALLY OCCURS FOR ONLY A FEW MINUTES. WHILE THIS TYPE OF TORNADO IS FAIRLY WEAK, IT CAN STILL LEAD TO PROPERTY DAMAGE.

EXPLORE MORE

A GUSTNADO IS ANOTHER KIND OF NON-SUPERCELL TORNADO. IT'S A SWIRLING MASS OF DUST OR DEBRIS THAT WHIRLS NEAR THE GROUND WITH NO FUNNEL. THIS TYPE OF TORNADO OCCURS BY THE GUST FRONT, OR A LINE OF VERY GUSTY WINDS, OF A THUNDERSTORM.

GUSTNADOES LIKE THIS ONE DON'T LAST VERY LONG.

Tornadoes also have "cousins" that cause considerable damage too. Dust devils are "whirlwinds" that occur in the desert or other very dry areas. These winds swirl dust around but commonly aren't stronger than 70 miles (113 km) per hour.

Forest fires or volcanic eruptions can cause fire whirls. These firenadoes may have winds of more than 100 miles (161 km) per hour.

DUST DEVILS CAN VARY IN SIZE, RANGING FROM AROUND 10 FEET (3 M) TO 100 FEET (30 M) WIDE. THIS "WHIRLWIND" USUALLY REACHES A HEIGHT OF AROUND 650 FEET (198 M).

CONSIDERABLE:
LARGE IN SIZE OR QUANTITY

IT'S RARE FOR FIRE WHIRLS TO LAST LONG ENOUGH FOR SOMEONE TO TAKE A PICTURE. THIS PHOTO OF A FIRE WHIRL WAS TAKEN NEAR SANTA BARBARA, CALIFORNIA, IN JUNE 2016.

TRACKING TORNADOES

Tornadoes have been reported every day of the year and at every time of the day. It's easy to understand why **predicting** them is so difficult. However, **meteorologists** keep track of the most common times and dates that tornadoes occur. Most tornadoes occur in April, May, and June. They happen most often in the late afternoon and early evening.

SKYWARN MEMBERS ARE OFTEN POLICE OFFICERS AND FIREFIGHTERS WHO HAVE BEEN TRAINED TO SPOT TORNADOES.

VOLUNTEER:
A PERSON WHO WORKS WITHOUT BEING PAID

EXPLORE
MORE

SKYWARN IS THE NAME OF A GROUP OF PEOPLE ALL OVER THE UNITED STATES TRAINED TO WATCH FOR DANGEROUS WEATHER SUCH AS TORNADOES. IF THEY SEE A FUNNEL CLOUD FORMING, THEY CALL A SPECIAL TELEPHONE NUMBER.

SKYWARN HAS BETWEEN 350,000 AND 400,000 VOLUNTEER "SPOTTERS."

Today, meteorologists know a lot about tornadoes. Some scientists are even experts on the subject. However, many years ago, they didn't have a clear idea about why funnel clouds formed, so it was hard for them to know when a tornado might hit.

EXPERT:
SOMEONE WHO KNOWS A GREAT DEAL ABOUT SOMETHING

WHILE THERE IS A LOT MORE KNOWLEDGE TODAY ABOUT HOW A TORNADO FORMS, IT IS STILL A DIFFICULT DISASTER FOR SCIENTISTS TO PREDICT EXACTLY. IT IS EASIER TO PREDICT THE ARRIVAL OF A HURRICANE THAN A TORNADO!

EXPLORE
MORE

THERE ARE ROUGHLY AROUND 100,000 THUNDERSTORMS A YEAR IN THE UNITED STATES. OUT OF THOSE THUNDERSTORMS, ABOUT 10 PERCENT OF THEM WILL TURN INTO SEVERE THUNDERSTORMS. AND ONLY AROUND 5 TO 10 PERCENT OF THOSE SEVERE THUNDERSTORMS WILL CREATE A TORNADO.

In order to predict a tornado today, meteorologists use radar and **satellites** to watch winds all over the country. They find areas in which conditions might cause thunderstorms. Doppler radar allows forecasters to estimate wind speeds and watch for mesocyclones. When they see the spinning winds reach a certain formation or speed, they issue a tornado watch.

A PROFESSIONAL STORM CHASER LOOKS AT RADAR ON HIS SMARTPHONE IN ORDER TO TRACK A SUPERCELL THUNDERSTORM.

ESTIMATE:
TO MAKE A CAREFUL GUESS ABOUT AN ANSWER BASED ON THE KNOWN FACTS

DOPPLER RADAR

The National Weather Service uses Doppler radar to forecast the weather. "Radar" stands for RAdio Detection And Ranging. Radar machines were first used to spot aircraft during World War II. Radar sends out radio waves that reflect off things like raindrops. The waves echo back and reveal information about the size, shape, and location of objects. Using many of these echoes over a period of time, meteorologists can find out how fast a storm is moving. Radar can also monitor different types of **precipitation** and which way clouds are turning, as well as wind strength.

NATIONAL DOPPLER RADAR SITES

ALASKA

HAWAII

GUAM

PUERTO RICO

● NWS/DOD RADAR SITES

THIS MAP SHOWS THE LOCATIONS OF THE VARIOUS DOPPLER RADAR SITES RUN BY THE NATIONAL OCEANIC AND ATMOSPHERIC ADMINISTRATION (NOAA) THROUGHOUT THE UNITED STATES.

In the 1970s, the United States started rating tornado damage from 0 to 5 on the Fujita Scale, or F Scale. This scale also uses the damage a tornado causes to estimate wind speed.

In 2007, meteorologists adopted the Enhanced Fujita Scale, or EF Scale. It gives more damage indicators than the F Scale and takes into account different kinds and amounts of damage. For example, if the outer walls of a small home fall because of a tornado, damage indicators estimate that the winds were about 132 miles (212 km) per hour. Based on this one indicator, the tornado would rate a 2 on the EF Scale.

AN EF-5 TORNADO CAN PULL DOWN STRONG BUILDINGS AND PICK UP AND THROW CARS AND TRUCKS!

ENHANCED FUJITA SCALE (EF SCALE)

⚠ EF RATING	3-SECOND GUST	DAMAGE
EF-0	65–85 MILES (105–137 KM) PER HOUR	LIGHT
EF-1	86–110 MILES (138–177 KM) PER HOUR	MODERATE
EF-2	111–135 MILES (178–217 KM) PER HOUR	CONSIDERABLE
EF-3	136–165 MILES (218–266 KM) PER HOUR	SEVERE
EF-4	166–200 MILES (267–322 KM) PER HOUR	DEVASTATING
EF-5	OVER 200 MILES (322 KM) PER HOUR	INCREDIBLE

SCIENTISTS ESTIMATE HOW FAST WINDS ARE BLOWING BY OBSERVING HOW MUCH DAMAGE IS CAUSED BY THE WINDS. SCIENTISTS LOOK AT THE DAMAGE OF A NUMBER OF BUILDINGS OR PARTS OF NATURE TO DETERMINE AN EF RATING.

VORTEX—or Verification of the Origins of Rotation in Tornadoes EXperiment—was a large tornado research project from 1994 to 1995 that explored how, when, and why tornadoes form. However, some of the facts scientists collected just produced more questions.

THIS PHOTOGRAPH FROM WESTERN NEBRASKA SHOWS A RADAR TRUCK USED DURING VORTEX 2'S STUDIES. THIS RADAR TRACK WAS COMPLETING SCANS DURING A THUNDERSTORM.

VORTEX NEXT STEPS

In 2009, VORTEX 2 became the largest tornado study ever. About 100 scientists used the latest weather instruments to find and track tornadoes. They had trucks outfitted with radar called Doppler on Wheels and tornado pods, which measure wind speed and direction on the ground. One of VORTEX 2's major targets was figuring out how to give communities more warning about tornadoes headed in their direction. VORTEX-SE then launched in 2016. Again, a large group of scientists gathered to examine the storms that cause tornadoes in the southeastern United States. The project is still ongoing.

DOPPLER ON WHEELS

TARGET:
THE FOCUS OF EFFORT

TORNADO TRAINING

A tornado can be measured by the width of its funnel, the speed of its winds, or the damage it causes. The deadliest tornado in U.S. history was the Tri-State Tornado of 1925. It formed March 18, 1925, in Ellington, Missouri.

The Tri-State Tornado traveled 219 miles (352 km) through Missouri, Illinois, and Indiana. The winds spun at 300 miles (483 km) per hour, destroying several towns and leaving many people homeless. During its 3½-hour lifespan, the tornado caused 695 deaths and more than 2,000 injuries. Although the tornado happened before the EF Scale was put in place, meteorologists believe it was an EF-5 rated tornado.

THE TRI-STATE TORNADO HAD THE LONGEST RECORDED TORNADO LIFESPAN IN HISTORY. THIS PHOTOGRAPH SHOWS THE REMAINS OF WEST FRANKFORT, ILLINOIS, FOLLOWING THE DESTRUCTION OF THE TRI-STATE TORNADO.

LIFESPAN:
HOW LONG SOMETHING IS ACTIVE

MAY 2019 TORNADO OUTBREAK

In late May 2019, a tornado outbreak with record numbers raged through the central United States. During May, there were 555 tornadoes reported. On average, 276 tornadoes are usually reported in May in the United States. The bulk of the tornadoes were reported in three outbreaks. From May 17 to 18, 67 tornadoes touched down, mainly in western Nebraska and Kansas. Then, from May 20 to 22, 119 tornadoes moved through Texas, Oklahoma, and Missouri, leaving considerable damage in their wake. Finally, from May 26 to 29, 190 tornadoes devastated areas from Colorado to Pennsylvania. This high number of tornadoes was responsible for property damage, several deaths, and injuries to over a hundred people.

THIS PHOTOGRAPH WAS TAKEN ON MAY 28, 2019, IN DAYTON, OHIO. IT SHOWS SOME OF THE DESTRUCTION FROM ONE OF THE DEVASTATING TORNADOS DURING THE 2019 OUTBREAK.

FOLLOWING THE 2011 JOPLIN TORNADO, THE FEDERAL EMERGENCY MANAGEMENT AGENCY (FEMA) WAS CALLED IN TO HELP CLEAN UP AND RESCUE THOSE TRAPPED IN FALLEN BUILDINGS.

EXPLORE MORE

ANOTHER DEVASTATING EF-5 RATED TORNADO WAS THE 2011 JOPLIN TORNADO. IT WAS THE WORST U.S. TORNADO IN MORE THAN 50 YEARS TO HIT A CITY. MORE THAN 150 PEOPLE DIED AND THE DAMAGE WAS WIDESPREAD. SOME BUILDINGS WERE RIPPED RIGHT OFF THE GROUND.

A tornado outbreak is the occurrence of many tornadoes in an area, commonly as a result of a thunderstorm system. The Tornado Super Outbreak of 2011 was the largest in U.S. history. From April 26 to April 28, more than 300 tornadoes spun across the eastern United States. In the 15 states hit by tornadoes, about 340 people were killed. More than 2,000 people were injured in Alabama alone.

ABOUT TWO WEEKS BEFORE THIS SUPER OUTBREAK, 155 TORNADOES WERE REPORTED IN SEVERAL SOUTHERN STATES BETWEEN APRIL 14 AND 16, 2011. THIRTY-EIGHT PEOPLE DIED DURING THIS OUTBREAK.

THE EFFECTS OF CLIMATE CHANGE

Climate change is the long-term change in Earth's climate, caused partly by human activities such as burning oil and natural gas. This has directly led to an increase in global temperatures. In particular, climate change has led to ocean surface temperatures increasing. This, in turn, has led to more moisture flowing into Earth's atmosphere. This increase of warm, moist air in the atmosphere may lead to more thunderstorms and tornadoes. But as tornadoes are still fairly **unpredictable**, there isn't a lot of direct evidence that climate change will lead to more tornadoes. However, scientists have seen an increase in tornado outbreaks, as well as a shift of Tornado Alley toward the Southeast. Scientists believe both may be related to climate change.

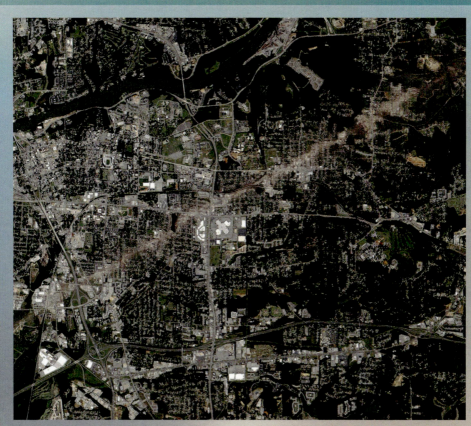

TUSCALOOSA, ALABAMA
APRIL 29, 2011

EVIDENCE:
SOMETHING THAT HELPS SHOW OR DISPROVE THE TRUTH OF SOMETHING

The United States has an average of more than 1,000 tornadoes reported every year. For American citizens, it's important to know how to stay safe during a tornado.

The safest place to be during a tornado is in your home's basement. If your home doesn't have a basement, crouch under a sturdy table or take shelter in a room with thick walls and no windows, such as a closet or bathroom. Cover your head with your arms to protect yourself from flying glass or debris.

IN SOME PARTS OF THE UNITED STATES WHERE TORNADOES ARE COMMON, PEOPLE MAY HAVE TORNADO SHELTERS NEAR THEIR HOMES OR IN THEIR TOWNS.

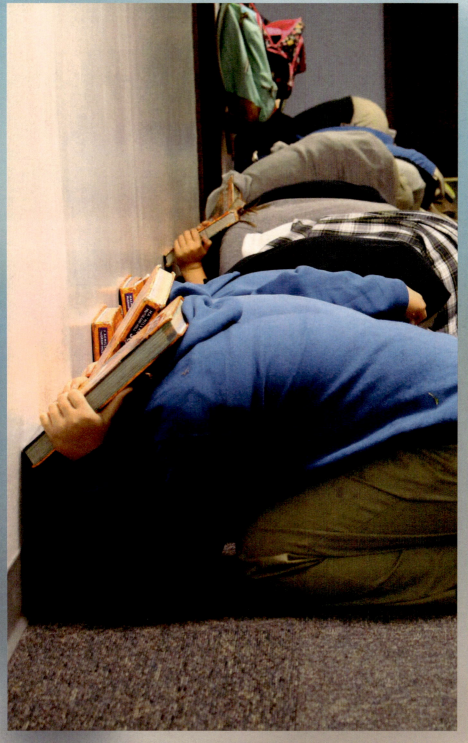

FLYING DEBRIS IS ONE OF THE BIGGEST DANGERS DURING A TORNADO. IF YOU'RE INSIDE WHEN A TORNADO STRIKES, IT'S IMPORTANT TO STAY AWAY FROM WINDOWS AND PROTECT YOUR HEAD AND NECK USING YOUR ARMS, LIKE THE STUDENTS PICTURED HERE.

If you're outside when a tornado hits, find an area of low ground, such as a ditch. Don't stay inside a car or mobile home because tornadoes can be strong enough to turn these over or lift them off the ground.

While tornadoes are a dangerous reality in the United States, technology has developed so people are warned earlier of dangers and risks. While the dangers of a tornado shouldn't be underestimated, there are ways people can prepare and stay safe in a tornado.

WHILE TORNADOES ARE A SIGHT TO SEE, THEY ARE ALSO UNPREDICTABLE AND POSE MANY THREATS.

REALITY:
THE TRUE STATE OF MATTERS

EXPLORE MORE

IS A TORNADO ABOUT TO HIT? THE SKY GIVES SOME CLUES. IT BECOMES DARK AND A GREENISH COLOR. A WALL CLOUD MAY FORM SUDDENLY AND BEGIN TO SPIN. IT MAY START TO HAIL. YOU MIGHT ALSO HEAR A LOUD ROAR, LIKE A TRAIN IS APPROACHING.

YOU CAN STAY PREPARED BY KNOWING YOUR AREA'S WEATHER CONDITIONS AND WHERE YOU CAN SEEK SHELTER.

45

GLOSSARY

condense To lose heat and change from a gas into a liquid.

damage Harm; also, to cause harm.

debris The remains of something that has been broken down.

front The line between two masses of air.

horizontal Level with the line that seems to form where Earth meets the sky.

meteorologist Someone who studies weather, climate, and the atmosphere.

precipitation Rain, snow, sleet, or hail.

predict To guess what will happen in the future based on facts or knowledge.

rotate To turn around a fixed point.

satellite An object that circles Earth in order to collect and send information or aid in communication.

unpredictable Not able to be known before happening.

vertically In an up and down manner.

FOR MORE INFORMATION

BOOKS

Challoner, Jack. *Eyewitness: Hurricane & Tornado.* New York, NY: DK Publishing, 2014.

Kostigen, Thomas. *Extreme Weather: Surviving Tornadoes, Sandstorms, Hailstorms, Blizzards, Hurricanes, and More!* Washington, DC: National Geographic, 2014.

Ruckman, Ivy. *Night of the Twisters.* New York, NY: Crowell, 1984.

Tarshis, Lauren. *I Survived the Joplin Tornado, 2011.* New York, NY: Scholastic Inc., 2015.

WEBSITES

How Do Tornadoes Form?
www.britannica.com/story/how-do-tornadoes-form
Watch a short video about how tornadoes come to be at this site.

Tornadoes Impact Our Lives
eo.ucar.edu/kids/dangerwx/tornado1.htm
Learn more about tornadoes and where the greatest number of tornadoes form.

Tornado Facts!
www.natgeokids.com/uk/discover/geography/physical-geography/tornado-facts
Learn more facts about tornadoes here.

INDEX

C

climate change, 41
condensation funnel, 16
cyclonic/anticyclonic tornadoes, 18

D

damage/destruction, 4, 6, 11, 19, 21, 24, 32, 36, 38, 39
deaths, 4, 36, 38, 39, 40
Doppler radar, 30, 31, 35
dust devils, 24

E

Enhanced Fujita Scale, 32, 36, 39

F

fire whirls, 24
forecasting/predicting tornadoes, 7, 26, 30, 31, 41
formation of tornadoes, 8, 9, 10, 12, 14, 20, 21, 22, 34

Fujita Scale, 32
funnel clouds, 16, 27, 28

G

gustnadoes, 23

H

hurricanes, 14

J

Joplin Tornado, 39

L

landspouts, 22
long-track tornadoes, 21

M

mesocyclones, 12, 30

N

National Weather Service, 7, 31

S

Skywarn, 27
storm chasers, 13

supercell thunderstorms, 9, 20, 22
supercell tornadoes, 20, 22

T

thunderstorms, 8, 9, 20, 22, 23, 29, 30, 40, 41
Tornado Alley, 11, 41
tornado outbreaks, 38, 40, 41
Tornado Super Outbreak, 2011, 40
tornado warnings, 7
tornado watches, 30
Tri-State Tornado, 36

V

VORTEX, 34
VORTEX 2, 35
VORTEX-SE, 35

W

wall clouds, 12, 45
warning signs of a tornado, 45
waterspouts, 22